Ibrahima Ouibga

Agricultural entrepreneurship in schools: the case of Bassemyam CEG

Ibrahima Ouibga

Agricultural entrepreneurship in schools: the case of Bassemyam CEG

ScienciaScripts

Cover image: www.ingimage.com

This book is a translation from the original published under ISBN 978-620-6-72327-1.

Publisher:
Sciencia Scripts
is a trademark of
Dodo Books Indian Ocean Ltd. and OmniScriptum S.R.L publishing group

120 High Road, East Finchley, London, N2 9ED, United Kingdom
Str. Armeneasca 28/1, office 1, Chisinau MD-2012, Republic of Moldova, Europe
Printed at: see last page
ISBN: 978-620-8-32744-6

DEDICATION

To our wife Samiratou KABORE,

For her infinite love and invaluable support.

ACKNOWLEDGEMENTS

We would like to express our gratitude:

- to Dr. Dieudonné TIBIRI for supervising this Master's thesis, and to all the teachers at the Institut International de Management de Ouagadougou for the training we received;
- the staff of the Swiss Cooperation Office/Burkina Faso for the various co-financed projects;
- to Mr. Panoussa COMPAORE for having pleaded the cause of the Collège d'Enseignement Général (CEG) de *Bassemyam* with the Swiss Cooperation;
- Mr. Damas ZOURE for his role as consultant in the entrepreneurship development project at the CEG de *Bassemyam*;
- to the staff and students of CEG de *Bassemyam*, for their involvement;
- to you Jules WIBGA, to Mrs Véronique OUIBGA née SIMPORE and to Boureima OUIBGA, we thank you for your support and encouragement;
- to our sister Awa KARAMBIRI née OUIBGA for her moral support;
- to Olivier A. OUIBGA, Adama SANFO and our other brothers and sisters;
- to our friends Emmanuel W. SAVADOGO, Boukaré KIEMA, Badiel Roland, and Amidou KABORE for their guidance, motivation, hard work and invaluable support.

FOREWORD

To obtain a professional Master's degree, students must submit a dissertation to a jury set up within the Institut International de Management. It is a requirement of the French Ministry of Higher Education, Research and Innovation to defend an intellectual work before a jury. In keeping with this tradition, we have chosen the theme of : "Contribution de la Coopération Suisse dans la promotion de l'entrepreneuriat agricole au CEG de *Bassemyam* dans la commune rurale de *Komsilga*". The choice of this theme reflects our desire to reflect on the promotion of agricultural entrepreneurship among young people.

To carry out this work, we began by collecting documents relating to agricultural entrepreneurship that might be of interest to us (dissertations, general works, articles, laws, decrees, etc.). We then carried out fieldwork, collecting data through interviews with managers and a questionnaire sent to staff who could provide us with information. This provided us with the information we needed to analyze the data collected. Finally, we wrote our dissertation after analyzing the data collected and analysed, with a view to making our contribution to promoting agricultural entrepreneurship in schools and rural areas.

LIST OF ACRONYMS AND ABBREVIATIONS

APE : Parents' Association

CEG : Collège d'Enseignement Général

CNRTL : Centre National des Ressources Textuelles et Lexicales **CNS-BF: Conseil**

National du Burkina **Faso** Conseil National de la Statistique du Burkina Faso **CVD** :

Comité Villageois de Développement (Village Development Committee)

SDC Swiss Agency for Development and Cooperation

EPS Physical and Sports Educator

FAO Food and Agriculture Organization of the United Nations

IFFAfrique Institut de Formation Fondacio Afrique

INSEE National Institute of Statistics and Economic Studies

MDAEP Ministry of Development, Economic Analysis and Forecasting

OFE Observatoire de la Francophonie Economique

ILO International Labour Organization

NGO non-governmental organization

PAPEA Programme d'Appui à la Promotion de l'Entrepreneuriat Agricole (Support program for the promotion of agricultural entrepreneurship)

UNDP United Nations Development Programme

PPEA: Agricultural Entrepreneurship Promotion Project

RGPH General Population and Housing Census

SECO State Secretariat for Economic Affairs

SNV Stichting Nederlandse Vrijwilligers/Foundation des Volontaires Néerlandais

SUMMARY

The study of the Swiss Cooperation's contribution to the promotion of agricultural entrepreneurship is part of its focus on supporting students and rural communities in Burkina Faso in the fields of education, training and local development. What then are the main initiatives undertaken by Swiss Cooperation to promote agricultural entrepreneurship within the CEG of a rural commune?

? The aim of this study is to assess the effectiveness of Swiss Cooperation in promoting agricultural entrepreneurship in the rural commune of *Komsilga*, particularly at the CEG de *Bassemyam*. The aim of this study is to identify the specific measures taken by Swiss Cooperation to support agricultural entrepreneurship, and also to examine the impact of this Cooperation's results on students, on the one hand, and local communities, on the other. The results of the field investigations showed that poverty and unemployment are a brake on the expansion of populations, particularly young people, hence our main hypothesis, which states that the activities carried out by Swiss Cooperation in the rural commune of *Komsilga* have contributed to the promotion of agricultural entrepreneurship at the CEG de *Bassemyam* and to the capacity-building of local communities.

Key words: Cooperation-Promotion-Entrepreneurship-Agriculture.

ABSTRACT

The study of the Contribution of Swiss Cooperation to the promotion of agricultural entrepreneurship is part of highlighting the support for students and rural communities in Burkina Faso in the sectors of education, training and local development. What are the main initiatives undertaken by Swiss Cooperation to promote the growth of agricultural entrepreneurship within the CEG of a rural municipality? The objective of the study carried out aims to evaluate the effectiveness of Swiss Cooperation in promoting agricultural entrepreneurship in the rural commune of Komsilga, particularly at the CEG of Bassemyam. The ambition of this study is to identify the specific measures taken by Swiss Cooperation to support agricultural entrepreneurship, but also to examine the impact of the results of this Cooperation on students, on the one hand, and local authorities, on the other hand. The results of field investigations made it possible to see that poverty and unemployment are a brake on the expansion of populations, particularly young people, hence our main hypothesis which states that the activities carried out by Swiss Cooperation in the rural commune of Komsilga contributed to the promotion of agricultural entrepreneurship at the CEG of Bassemyam and to the capacity building of local communities.

Keywords: Cooperation-Promotion-Entrepreneurship-Agricultural.

CONTENTS

INTRODUCTION

A Sudano-Sahelian West African country, Burkina Faso, despite the onset of a demographic transition, remains a country with a high birth rate of 39.4, reflecting the rapid growth of its population (RGPH, 2019). This state of affairs, like a sea serpent, poses acute problems such as food security, nutritional quality and the employability of the population, especially the young. However, food security, nutritional quality and employability are permanent development objectives whose attainment in Burkina Faso remains severely hampered by the endemic combination of several factors. The corollaries of this situation include rising unemployment and food imbalance.

According to the final report of the national school health, hygiene and nutrition program of the Basic Education Strategic Development Program PDSEB (2017, p. 25), nutritional imbalance is nowadays a public health problem that affects the country's development policies. Indeed, 7.9% of schoolchildren suffer from acute malnutrition or wasting, 12% from chronic malnutrition and finally 11.7% from underweight. Malnutrition therefore threatens the young age bracket, particularly the under-15s who represent 48% of the population, and is the cause of many school problems including mainly nutritional deficiency-related illnesses, absenteeism, early school dropout, poor results according to the PDSEB report (2017, *op.cit.*, p. 20).

Satisfactory nutrition and education are therefore essential for children's development and future livelihoods. These priorities are reflected in the first and second Millennium Development Goals. However, for millions of children, these goals are far from being achieved. Children who go to school hungry cannot learn well. Their physical activity, intellectual capacity and resistance to infection are diminished. Yet the country's economy relies heavily on agriculture. But this main activity is still unable to adequately cover the population's food security needs and generate a type of agricultural wage-earning activity. This situation is attributable to natural hazards, outdated farming techniques and methods, and above all to the low level of development of agricultural entrepreneurship.

To curb the impact of this now nationwide problem, the Ministry of National Education, Literacy and Promotion of National Languages, in addition to its primary vocation of instructing, training and educating, is committed to promoting extracurricular activities such as school gardens. For general education as it stands today, with its long training cycles, poses the acute problem of the employability of learners. Pending the actual implementation and popularization of the reforms undertaken, and faced with the problem of unemployment which has become recurrent, it is appropriate, according to the current situation, to introduce learners to agricultural entrepreneurship and financial education, with a view to encouraging them to become self-employed, while helping to fix them in their communities.

To carry out our reflection on the contribution of Swiss Cooperation in the promotion of agricultural entrepreneurship within the CEG of *Bassemyam* in the rural commune of *Komsilga*, we have oriented our work in two essential parts of two chapters per part. The first part is entitled "Theoretical framework and literature review". In this part, we will, on the one hand, take a definitional approach to the essential notions in order to facilitate understanding of our subject. On the other hand, we will review the literature to address the works that have abounded on the issue of cooperation and agricultural entrepreneurship in rural areas. We will also highlight the institutional framework, presenting the Swiss Cooperation, the *Bassemyam* CEG and the study area. The second part of our paper is entitled "Research methodology and analysis of results". This second part deals with the methodological and analytical framework of the study. The methodological framework specifies the scientific approach adopted for data collection, and the analytical framework enables us to analyze and interpret the results.

PART ONE: THEORETICAL FRAMEWORK AND LITERATURE REVIEW

The first part of our study is devoted to a presentation of the research questions, objectives and hypotheses, a definitional approach to the key concepts of the subject and a review of the literature. The definitional approach enables us to provide an explanation of the essential terms of the subject (cooperation, promotion, entrepreneurship and agricultural entrepreneurship, rural commune) in order to better define them. This makes them easier to understand and avoids any confusion. In addition, a literature review will be carried out to see the various writings related to our research topic.

We focus on authors and writings related firstly to development cooperation, secondly to the promotion of agricultural entrepreneurship, and thirdly to agricultural entrepreneurship and its contribution to rural development. In this section, we will provide a general overview of the Swiss Cooperation, the CEG of *Bassemyam* and the study area. These two chapters will enable us to situate our research, mark out the ground for understanding and highlight our quest: that of the Swiss Cooperation's contribution to the promotion of agricultural entrepreneurship at the CEG de *Bassemyam*, in the rural commune of *Komsilga.*

CHAPTER I
THEORETICAL FRAMEWORK AND LITERATURE REVIEW

The theoretical framework concerns the research questions, objectives and hypotheses that are the prerequisites for our study. The analysis of concepts provides an explanation of the essential terms of our research subject. For us, it's a question of taking a definitional approach, providing elements of understanding for the key concepts of the subject. The central terms that we will explain will facilitate the understanding of our work and highlight our research objective. In addition, clarifying key concepts will help us to better define our quest and understand our study. The concepts to be defined are: cooperation, promotion, entrepreneurship and agricultural entrepreneurship, rural commune. Following these elements, we present the two institutions involved in our work (the institutional framework), as well as *Bassemyam*, the study area.

I.1. THE THEORETICAL FRAMEWORK

Promoting agricultural entrepreneurship in a post-primary school like the CEG de *Bassemyam* is an initiative that can help reduce the constraints (length of cycles, early school dropouts and failure, malnutrition, lack of motivation among learners, etc.) that are generally hampering the development of Burkina Faso's schools. So how can these constraints be reduced at a school like Collège d'Enseignement Général (CEG) de *Bassemyam*?

1.1.1. Problematizing the study

The search for solutions to the difficulties raised above prompted members of the CEG de *Bassemyam* educational community to approach the Swiss Cooperation Office, which responded favorably to their approach by setting up a project to promote agricultural entrepreneurship. The idea is to create a learning platform through a school farm. Then, to make the *Bassemyam* CEG a pilot school to introduce learners to entrepreneurship and financial education (Sol- Vert, 2022). The project involved students, teachers and parents in poultry farming, agroforestry and agro-ecology (see photo plates in appendices XVII and XVIII).

The project has enabled a strong collaboration between the educational community and local authorities. The CEG de *Bassemyam* has become an open-air laboratory where learners can learn about entrepreneurship in a concrete and experimental way. It is with this in mind that we are directing our inquiry into the contribution of Swiss cooperation to the CEG de *Bassemyam*, by asking the following question: What is the contribution of Swiss cooperation to the promotion of agricultural entrepreneurship at the CEG de *Bassemyam?* ? From this main question flow the following secondary questions: What are the main initiatives undertaken by Swiss Cooperation to foster the development of agricultural entrepreneurship within the CEG of a rural commune? Have the agricultural activities carried out by Swiss Cooperation contributed to improving the entrepreneurial skills of the beneficiaries? What are the objectives of this study?

1.1.2. Study objectives

The aim of this study is to analyze the contribution of Swiss Cooperation in promoting agricultural entrepreneurship in the rural commune of *Komsilga*, particularly at the CEG de *Bassemyam*. In addition, our study aims to identify the specific initiatives taken by Swiss Cooperation to promote agricultural entrepreneurship in the rural commune of *Komsilga*. In addition, we aim to study the impact of Swiss Cooperation actions on the entrepreneurial skills acquired by local communities, as well as on their economic and social well-being.

1.1.3. Study assumptions

In light of the research questions and objectives highlighted, we formulate the following hypotheses to guide our research and understand the involvement of Swiss Cooperation in the rural commune of *Komsilga*. Our main hypothesis states that the activities carried out by Swiss Cooperation in the rural commune of *Komsilga* have contributed to the promotion of agricultural entrepreneurship at the CEG de *Bassemyam* and to capacity-building in local communities. From this main hypothesis flow the following secondary hypotheses: Swiss Cooperation has facilitated learners' access to appropriate training, technical resources and market and income opportunities. Our second secondary hypothesis asserts that Swiss cooperation has contributed to improving beneficiaries' entrepreneurial skills and their economic and social well-being. It is useful to conceptualize the key terms of the study in order to remove any ambiguity as to their use throughout the analysis.

I.1.4. Study conceptualization

Defining the main concepts of the study is a fairly important exercise, which makes it possible to clarify the use of certain expressions specific to our study context. The aim of this section is to put potential readers on the same level of understanding with regard to the use of certain expressions whose meaning may be misunderstood or may vary from one region to another, depending on the vocabulary used. These expressions include: Cooperation, Promotion, Agricultural Entrepreneurship, Rural Commune.

I.1.4.1. Cooperation

The Centre National des Ressources Textuelles et Lexicales (CNRTL 2012) defines cooperation as the action of participating, with one or more people, in a common work or action. Indeed, cooperation is a kind of help, mutual aid or even agreement between the members of a group, an association with a view to a common goal. What's more, it's a method of action by which people with common interests join forces for profit. To this end, each member or stakeholder is invested with a responsibility, a mission and/or a right over the activities to be carried out.

Daouda Ouattara (2020, p. 8.) in his Master's Thesis, describes cooperation as:

"A state of mind and a mode of behavior in which individuals conduct their relations and exchanges in a non-confrontational or non-competitive way, seeking appropriate modalities for analyzing situations in a shared way and collaborating in the same spirit to achieve common or mutually acceptable ends. It is a policy of understanding and exchange between two countries".

For Marcel Pariat (2016), cooperation is a set of activities that enable us to enter into relationships with others. It fosters a climate of trust and mutual respect. In addition, it aims to develop skills and attitudes that foster, among other things, trust, dialogue and communication. Moreover, José Antonio Alonso and Jonathan Glennie (2015) see cooperation as an action for development.

To this end, they present cooperation as an asset for development. For them, cooperation takes place at three levels. Firstly, it involves financial transfers. Second, it is about capacity building. Thirdly, it emphasizes policy change. The capacity-building aspect is the one that fits into the dynamics of our thinking, along with the term promotion.

I.1.4.2. Promotion

The CNRTL (2012), defines promotion as the implementation of a project, the creation of a thing. It consists in provoking the development of a thing. It facilitates its success by carrying out a succession of activities. In the context of a project, promotion helps to raise visibility, to give the project greater prominence, to encourage its development and smooth running.

Fiches Pratiques Chef d'Entreprise (https://fiches-pratiques.chefdentreprise.com, 2016) defines promotion as the action of promoting something, an activity, an organization. Promotion is essential to the development of any structure. To this end, the action of promoting fosters the development of a multitude of strategies, the building of a brand image and notoriety, the development of communication, the expansion of partnerships, the enlargement of one's network, a human, material and financial investment.

In the context of our topic, promotion is linked to entrepreneurship, particularly in rural areas. The aim is to provide training and support on farms. In this respect, the Référentiel de formation " Entreprenariat Agricole ", *Guide du formateur en entreprenariat agricole* (Duddal, 2023), specifies that entrepreneurial promotion aims to fill gaps and effectively achieve objectives. It's a way of supporting farmers to manage their operations more rationally. This leads to greater profits and contributes to dynamism.

I.1.4.3. Agricultural entrepreneurship

Saiba Bakouan et al (2020) defines entrepreneurship as the act of undertaking, developing a project. Nassera Jouhari (2017) indicates that the term is regularly used to indicate the creation of an economic activity to meet a need, an expectation or for the achievement of a specific objective. In addition, the term entrepreneurship is very broad.

To this end, SumUp (https://www.sumup.com/en-en/business-guide/faq-familiarize-yourself-with-sum-up/2020) encompasses different types of entrepreneurship, including : ex-nihilo entrepreneurship (creating a new business), company takeovers (buying out a company), intrapreneurship (running an economic activity within the same company), extrapreneurship (employees of a company creating their own business), franchise entrepreneurship (setting up a business under a parent company), auto-entrepreneurship (self-employment as a framework for economic activity), public entrepreneurship (setting up public enterprises), social entrepreneurship (setting up companies carrying out economic activities of a social nature).

Social entrepreneurship is the one we're interested in. It is part of our topic of reflection on the contribution of Swiss cooperation to the promotion of agricultural entrepreneurship in a rural commune. Agricultural entrepreneurship in the rural commune of *Komsilga* is akin to social work. It's a business or even an agricultural creation initiative. Bobby Juuko (2022) in "Entrepreneuriat agricole" presents agricultural entrepreneurship as a key to developing the potential of rural areas on the African continent. Finally, IFFAfrique (2024), Institut de Formation Fondacio Afrique, evokes agricultural entrepreneurship in reference to both economic and social potential for rural populations.

I.1.4.5. Rural commune

The Institut National de la Statistique et des Etudes Economiques (INSEE 2022), defines communes as varying greatly in size. INSEE indicates that some may appear as sparsely populated or, on the contrary, densely populated, even though their populations are of comparable size. We thus speak of densely populated communes, intermediate density communes, sparsely populated communes and very sparsely populated communes. The first two (densely populated and intermediate density) correspond to the urban space, to urban communes. On the other hand, the last two (sparsely populated and very sparsely populated) refer to rural areas.

Xavier Browaeys and Paul Chatelain (2011), contrast the rural commune with the urban commune to better differentiate them. For them, the rural commune is more akin to the countryside, the village. It's a peaceful, less noisy environment, conducive to farming and livestock breeding. The urban commune, on the other hand, is seen as a city, an urban center, a conurbation. It is opposed to the former (rural commune) by its intense activity, density, population mobility, various infrastructures and modern roads.

The Conseil National de la Statistique du Burkina Faso (CNS-BF 2009) defines a rural commune as a group of villages with a population of at least 5,000, whose economic activities generate annual budgetary resources of at least five (5) million CFA francs. He added that the rural environment represents all localities that are not part of the urban environment. Finally, Moussa Ouédraogo et al (2009), in *Les capacités des communes rurales au Burkina Faso*, describes the rural commune as an area with high economic potential, but low economic income and development constraints.

I.2. INSTITUTIONAL FRAMEWORK

This section focuses on the presentation of the Swiss Cooperation and the CEG de *Bassemyam* located in the rural commune of *Komsilga*. We will look at this partner of the CEG (Swiss Cooperation) through its involvement in governance and democracy, its presence in basic education and vocational training, its contribution to local economic development and its activity in the rural commune of *Komsilga* through the CEG of *Bassemyam*.

I.2.1. General information on Swiss Cooperation

This sub-section looks at the approaches and partners of Swiss cooperation, and the case for such cooperation in terms of governance and democracy. It also shows the impact of this cooperation on basic education and vocational training. Finally, indicate its contribution to local economic development.

I.2.1.1. Swiss Cooperation's approach

Daouda Ouattara (2020, *op.cit.*, p. 70) tells us that relations between Switzerland and Burkina Faso are among the oldest partnerships the country has forged since 1961 (see organization chart, page 13). A bilateral agreement on trade, investment protection and cooperation was signed in 1969, and was revised in 1978, followed by agreements on technical cooperation and budgetary aid in 2006 and 2009, for greater efficiency.

In 1974, Switzerland strengthened its presence in Burkina Faso with the opening of the Swiss Cooperation Office in Ouagadougou (see organization chart for Switzerland in Burkina Faso in Appendix X). From 1976 onwards, the Swiss Cooperation became involved in supporting farmers and livestock breeders, as well as micro-entrepreneurs. Since 1994, the State Secretariat for Economic Affairs (SECO) has been providing general budgetary support and technical assistance for the Burkinabe government's economic efforts. Cooperation has been uninterrupted ever since. To meet the country's challenges effectively, Switzerland combines its humanitarian interventions, development programs and commitment to peacebuilding and violence prevention in a coherent manner (SDC, 2021).

Swiss cooperation is therefore part of a dynamic of support for local populations. It supports them as part of a national process of supporting public policies that have an impact on their environment and living conditions (SDC, 2021). This is why it provides humanitarian aid to the most vulnerable populations (SDC, 2021). Indeed, since 2019, with the context of insecurity prevailing in Burkina, Swiss Cooperation has been committed, alongside the Burkinabe government, to facilitating conflict resolution, maintaining development gains, and reducing the difficulties faced by populations through farming, particularly agricultural entrepreneurship (Actualités locales, 2022). It also works to support Burkina Faso in various sectors, including governance, education, training, culture and local development.

I.2.1.2. Swiss Cooperation in Governance and Democracy

Swiss Cooperation works to strengthen the skills of local players, focusing on issues of governance, decentralization and citizenship. It supports the development and management of market, water and access infrastructures, and encourages the emergence of sustainable, high-quality public services.

Over the past four years, its support has enabled the construction of two hundred and twenty-one (221) large-scale infrastructures, benefiting nearly two (2) million people and stimulating the development of a local workforce, including twenty thousand (20,000) people for the construction of rural tracks. Swiss Cooperation also promotes dialogue between civil society and the authorities. It supports the media, including almost twenty-five (25) community radio stations covering 75% of the territory. This support for the media helps to strengthen the production and dissemination of reliable, quality information (SDC, 2021).

I.2.1.3. Swiss Cooperation for Education and Training

The Swiss Cooperation is at the service of basic education and vocational training, working to ensure a quality education and training system accessible to all. To this end, it contributes through various offers to improving the quality of education. In this field, Swiss Cooperation is distinguished by :

- teacher training ;
- the creation of teaching materials ;
- alternatives to the formal school system for children and adults who have not had access to education;
- vocational training in collaboration with professional associations, the private sector and regional public services;
- educational continuity for children displaced by insecurity.

In four years, this cooperation has enabled fifteen thousand (15,000) school dropouts, 50% of them girls, to access basic education and vocational training in agro-sylvo-pastoral trades. Also, it is important to point out that, since 2017, the commitment of Swiss Cooperation has enabled around eight hundred and seventy-five thousand (875,000) children who fled their villages for reasons of insecurity to attend school in a protective environment (SDC, 2021).

I.2.1.4. Swiss cooperation for local economic development

Swiss cooperation has helped to strengthen the local economy through actions that promote food security. To this end, it has built numerous community infrastructures. An example is shown in photo 1 above, at the CEG de *Bassemyam*.

<u>Photo no. 1</u>: Swiss Cooperation builds infrastructure at the CEG de *Bassemyam*

Source: OUIBGA Ibrahima, December 2023.

The same applies to facilitating employment, self-employment and income for rural populations. Indeed, to improve farmers' incomes and create jobs (especially for young people and women), Swiss Cooperation supports the modernization of family farms in various sectors:

- RICE GROWING ;
- market gardening ;
- non-timber forest products ;
- Livestock and livestock product sectors.

Swiss cooperation is based on the fact that better production, processing and marketing of agricultural products improves the profitability of family farms and, in turn, enhances food security. To this end, Swiss Cooperation support has improved the livelihoods of almost six hundred and twenty-five thousand (625,000) working people, 60% of whom are women, in just four (4) years.

In short, Swiss Cooperation's support for farmers' organizations strengthens their negotiating skills and enables rural producers to influence national policies (SDC, 2021). What can we say about the CEG de *Bassemyam*, a beneficiary institution committed to promoting agricultural entrepreneurship?

I.2.2. General information about CEG *Bassemyam*

The CEG, Collège d'Enseignement Général, which promotes agricultural entrepreneurship, bears the local name: CEG de *Bassemyam*. Like the other CEGs in Burkina Faso, the *Bassemyam* CEG has an administrative director and a teaching staff. The teaching staff is made up of teachers of the following disciplines: French, English, history-geography, mathematics, physical sciences and EPS (Educateur Physique et Sportif).

What's special about Burkina Faso's CEGs is that they are essentially made up of first-cycle classes: sixth, fifth, fourth and third. This represents a young, motivated and eager-to-learn population (A. Zongo, 2021). The vision of Swiss Cooperation at the CEG de *Bassemyam*, in the Commune Rurale *de Komsilga*, is to ensure that each CEG pupil is able to satisfy his or her essential needs and develop the self-esteem necessary for personal fulfilment through agricultural activities (see photo no. 2, p. 16).

Through the promotion of agricultural entrepreneurship, the aim is to contribute to a sustainable improvement in living conditions, and in turn, reduce poverty and unemployment among young people, particularly in rural areas. According to A. Zongo (2021), the Swiss Cooperation project with the CEG de *Bassemyam* is part of the drive to support young people and local authorities in promoting local economic development through agricultural entrepreneurship.

Photo 2: Learners and teachers in the garden at CEG *Bassemyam*

Source: OUIBGA Ibrahima, December 2023.

I.2.3. General information on the study area

Bassemyam or *Bassiyam* is a town in the Central region of Burkina Faso. It is located in *the Komsilga* department, in the province of *Kadiogo*, some ten kilometers from the capital, on the *Ouagadougou-Saponé* axis (see map no.1, page 17). According to the Fichier des localités du 5ème Recensement Général de la Population et de l'Habitat RGPH (2019, p. 67), its population was estimated at 2,809.

Map 1: Location of *Bassemyam* in the rural commune of *Komsilga*

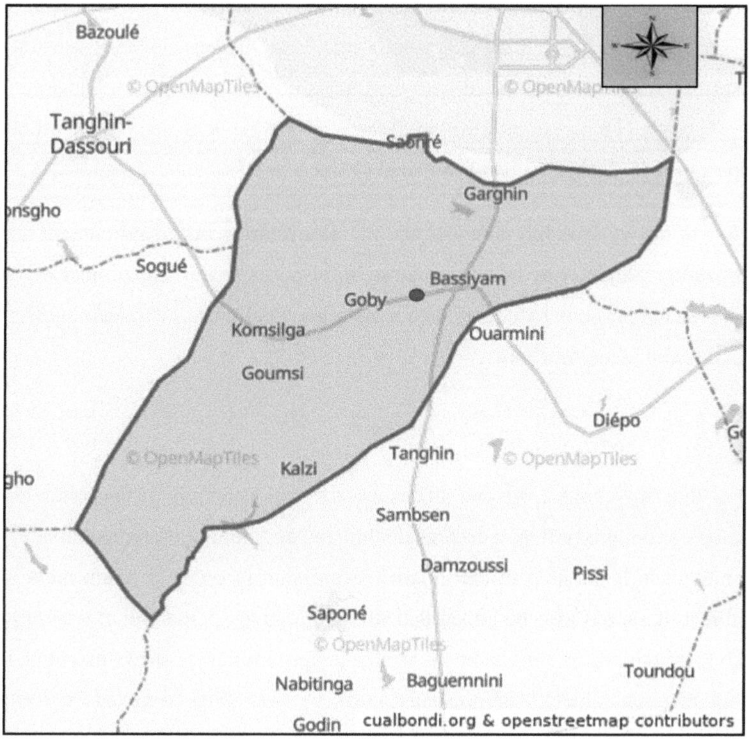

Caption:

⬬ : *Bassemyam*

▬ Commune rurale de *Komsilga* :

Source : https://cualbondi.org/bf/a/r5675754/komsilga/#google_vignette

Bassemyam's economic activity is based mainly on agropastoralism, aided by the irrigation provided by the nearby *Boulbi* dam. Bassemyam also boasts a well-known market in the area, which alternates with markets in *Kienfangué* and another village, and meets every three days in the same place.

However, the locality is strongly confronted with the repercussions of climate change, which are affecting its economic activities, including agriculture. The manifestations of these climatic changes are largely the consequence of man's bad activities on the environment, such as uncontrolled housing developments, abusive logging, degradation of riverbanks, etc. Maps n°2 and n°3 (in appendix XIX and XX respectively) illustrate the consequences of climatic events. Faced with the threats posed by rampant urbanization, we need to preserve the environment (the land) by involving young people in promoting farming activities that respect the environmental balance.

It was in view of this agricultural potential and the deteriorating rural environment that the Swiss Cooperation project was born. An agricultural entrepreneurship project involving young people, and in particular students with a more practical focus, with learners from the CEG de *Bassemyam* at the forefront.

*****

At the end of this first chapter, we can affirm that the understanding of the following key concepts: cooperation, promotion, entrepreneurship and agricultural entrepreneurship, rural commune, has made it possible to clarify each expression in order to relate them to the research subject. It should also be added that the definitional approach that was provided also helped to remove any confusion as to the understanding of these essential terms central to our research subject. The presentation of the institutions involved in promoting entrepreneurship in schools, namely the Swiss Cooperation and the CEG de *Bassemyam*, as well as the study area, was made in order to further identify the nature of their respective commitments to the success of the initiative.

CHAPTER II
REVIEW OF THE LITERATURE AND THE STATE OF THE QUESTION

A literature review is a critical assessment of research developments in a specialized field. In this section, devoted to expressing our knowledge of the research topic, we carry out a scientific literature search on the various writings and authors related to cooperation and the promotion of agricultural entrepreneurship in rural areas. For a better approach and realization of our reflection work, we consulted physical and digital documents: articles, reports, dissertations, books. Following the documentary survey, we carried out a critical analysis focusing on the actual state of the question in previous works concerning agricultural entrepreneurship.

II.1. REVIEW OF LITERATURE

The question of agricultural entrepreneurship has been addressed many times in the literature. To gain a better understanding of our research topic, we read both general and specific works, dissertations, articles and journals on agricultural entrepreneurship and initiatives to promote it.

II.1.1. Development cooperation

Josė Antonio Alonso and Jonathan Glennie (2015) address the issue of development cooperation in *Development Cooperation Forum Policy Notes*. Their discussion focuses on two strands: the objectives of development cooperation, on the one hand, and the criteria for development cooperation, on the other.

With regard to the objectives, they specify that it is important, firstly, to support and complement the efforts made by developing countries to ensure that essential and universal social standards are in place to enable citizens to exercise their fundamental rights. Secondly, to promote the convergence of developing countries towards higher levels of income and well-being, and reduce the profound inequalities that persist at international level. Finally, to support the active participation of developing countries in the provision of global public goods.

From the point of view of development cooperation criteria, Alonso and Glennie (2015), notify that development cooperation must, firstly, explicitly aim to support national or international development priorities. Secondly, it must not seek profit, but rather advocate positive discrimination in favor of developing countries. Finally, they believe that cooperation should be based on relationships that enhance ownership by developing countries.

Marcel Pariat (2016), in an article entitled "La coopération, une valse à trois temps?", published in *Pensée plurielle* magazine, points out that cooperation is important and necessary in terms of the contributions and innovation it orchestrates within structures, communities. Marcel Pariat sees cooperation as a waltz with three beats, or even three movements. To do this, he draws on three concepts: transaction (to compromise, to arrange, to make arrangements), transition (to improve the relationship between reform and development), transformation (to modify, to change, to move from one form to another). For him, cooperation cannot be thought of without these three elements. This enables us to analyze the relationships, resonances, interdependencies and context in which cooperation takes place, as well as its nature.

The University of Peace (2018) is more in line with the dynamic of providing guidelines for developing cooperation. In the publication entitled "Seven steps to developing cooperation", it shows that cooperation enables us to enter into relationships with others. Indeed, it fosters a climate of trust and mutual respect. The aim of cooperation is to develop skills and attitudes that foster trust, internal and external communication, and dialogue between members and between different parties. It's about experiencing and learning the spirit of mutual aid, collaboration and cooperation. In order to achieve this, the Université la Paix has identified seven key principles

(7) essential steps :

- **being welcomed**: feeling comfortable and safe in the group ;
- **express themselves**: develop individual expression ;
- **recognizing one's own qualities and those of others**: fostering a positive image between members ;
- **take your place and recognize that of others**: distribute roles and tasks;
- **listening**: using all the senses, learning to listen to others fully;
- **developing self-confidence and trust in others**: strengthening mutual trust ;
- **cooperating, building together**: experimenting with collective creations.

II.1.2. Promoting agricultural entrepreneurship

Saiba Bakouan et al (2020), in *Développement économique et emploi en Afrique francophone, l'entrepreneuriat comme moyen de réalisation*, a collective work by the Observatoire de la Francophonie Economique (OFE) in collaboration with the Université de Montréal, addresses the notion of entrepreneurship in Burkina Faso as a beacon of hope for an increasingly ambitious youth. Saiba Bakouan et al. take as their starting point an overview of the state of entrepreneurship in Burkina Faso, focusing on aspects relating to the labor market, forms of employment, the entrepreneurial spirit and the main challenges facing Burkina in conjunction with political decision-makers and the population, particularly young people. Saiba Bakouan et al. see self-employment (entrepreneurship) as a way out of unemployment. Indeed, self-employment offers a glimpse of hope, improved living conditions and a way out of precariousness.

HELVETAS Burkina Faso (2020), in its Programme d'Appui à la Promotion de l'Entrepreneuriat Agricole (PAPEA) in Burkina Faso, implemented in collaboration with the NGO SNV, Stichting Nederlandse Vrijwilligers/Foundation des Volontaires Néerlandais, promotes agricultural entrepreneurship through the direct and/or indirect creation of activities and jobs. By encouraging agricultural entrepreneurship, they enable the rural population, especially women and young people, to generate income. HELVETAS aims to revitalize the rural agricultural sector in the North, Centre-West, Hauts-Bassins and peri-urban areas of Bobo-Dioulasso and Ouagadougou. This revitalization involves processing and preserving locally-produced agricultural produce. This will strengthen sustainable and inclusive food systems in rural and peri-urban areas.

The project (HELVETAS Burkina Faso/PAPEA 2020) aims to set up 60 clusters, support 6,000 agricultural businesses, impact 24,000 people, including 14,400 young people and 12,000 women under 35, increase farm revenues by 10%, and generate 5,000 decent jobs for young people and women. As a reminder, the PAPEA began with the mango, manioc, ginger, poultry, potato, milk, egg, corn, peanut, honey, pork, tomato, strawberry, onion, cattle, tomato and pepper sectors.

With regard to the International Labor Organization (ILO 2022), the ProAgro project is helping to build institutional capacity to support agricultural entrepreneurship. The ProAgro project, in collaboration with Mohammed VI University, will support the design and implementation of the CRJEA, Centre Régional des Jeunes Entrepreneurs Agricoles et Agroalimentaires. The creation of this center aims to improve living conditions and encourage job creation, particularly for rural youth. The aim is to provide young people with technical assistance, training and local coaching. To this end, they have been dubbed "Generation Green". This name refers to a new generation of young agricultural entrepreneurs.

II.1.3. Agricultural entrepreneurship and rural development

The Projet de Promotion de l'Entreprenariat Agricole (PPEA 2015), in its *Activity Report to December 31, 2014,* asks about the "Socio-economic transformation of rural areas in Benin". The study carried out by the PPEA is the fruit of cooperation between the Beninese state and the United Nations Development Program (UNDP). It is under the supervision of the Ministry of Development, Economic Analysis and Forecasting (MDAEP), with technical support from the Songhaï regional center. The study is aimed at the socio-economic transformation of Benin's rural and agro-ecological zones. The project is the Beninese government's response to the triple challenge of youth employment, poverty and rural exodus.

In short, the PPEA works to :

- Ensure the development of a modern, sustainable, competitive agriculture focused on the promotion of agricultural value chains.
- Promote and disseminate innovative, high-performance agricultural technologies and practices.
- Ensure the development of human capital in Benin's various agro-ecological zones with a view to effectively transforming Benin's relative advantages into competitive advantages.
- Ensure the creation of economic opportunities at commune level by promoting inter-regional exchanges that facilitate the densification of local economies.

Bobby Juuko (2022), in *Entreprenante Afrique* magazine, discusses agricultural entrepreneurship as a key to developing the potential of rural Africa. Juuko's starting point is that the African continent has sufficient resources. Indeed, he points out that the Food and Agriculture Organization of the United Nations (FAO) estimates that unused arable land in Africa exceeds 100 million hectares.

He also highlights the fact that many countries on the continent still have a dominant rural population. This is an opportunity that Africa should take advantage of. However, Juuko points out that despite these assets, a number of challenges are preventing the agricultural sector from taking off. He refers to land policies, logistical problems, agricultural subsidies and lack of productivity gains. However, Juuko proposes that the solution lies in focusing on and satisfying the continent's internal demand, in order to stimulate the development of modern agriculture in Africa. In addition, he points out that increasing yields can bridge the gap with population growth. In addition, he adds, the exploration of fruit and vegetable crops as an increasingly attractive alternative.

II.2. THE CURRENT STATE OF AGRICULTURAL ENTREPRENEURSHIP EFFICIENCY ISSUES

Some authors question the efficiency of agricultural entrepreneurship. They believe that, in general, its implementation is the result of spontaneity, comparison or a rather constraining situation brought about by poverty and/or unemployment, and not the outcome of a rational and entrepreneurial path. However, based on the opinions of other authors as well as our own, we have put forward an argument to justify its relevance, as long as it conforms to and contextualizes the entrepreneurial approach that suits it.

II.2.1. Mixed opinions

Many authors are convinced of the futility of agricultural entrepreneurship, especially in the younger age groups, because despite support, the initiative can end in failure. According to Léger-Jarniou (2005), failure can be understood as the abandonment of the entrepreneurial project or the bankruptcy of the supported enterprise, or the ineffective and inefficient management of the supported enterprise.

Entrepreneurial failure is also defined as the entry of the entrepreneur into a psychological state of disappointment, which can lead to the demise of the business, when financial and/or moral support has not been forthcoming. Failure can also be explained by a governance deficit marked by managerial shortcomings, weak managerial commitment, or the entrepreneur's ignorance of the challenges. Finally, Martine Boutary et al. (2014) point out that entrepreneurs' knowledge construction is not adapted to the realization of their projects.

Fitz Koch et al, (2017) finds other reasons for the failure of agricultural entrepreneurship. The author holds the main people responsible, i.e. the entrepreneurs themselves. He points out that while agriculture is a vector for development, the low level of agricultural entrepreneurship in Africa can be explained by the predominance of necessity-based entrepreneurs, who often live in precarious conditions or seek production efficiency.

Raphaël Nkakleu et al, (2023,) define entrepreneurs out of necessity as people who engage in entrepreneurial activities because of family or personal constraints, in other words, because they lack alternatives for finding a job. Necessity entrepreneurship is the dominant form of entrepreneurship in Africa. This is why Raphael Nkakleu (2023) justifies the slow development of African economies by the fact that projects created by necessity entrepreneurs have a low success rate. Entrepreneurs out of necessity are usually young entrepreneurs who find themselves in situations of isolation (lack of professional, personal or social networks) that limit the possibilities of transforming their business project, particularly in the start-up phase. Should we then, in the name of the aforementioned arguments, reduce agricultural entrepreneurship to its sole character of constraint or failure?

II.2.2. Favorable opinions

Like any human initiative, agricultural entrepreneurship is a source of both success and failure. It all depends on how it is approached and, above all, implemented. That said, it's the absence of an entrepreneurial culture that is generally the source of setbacks. Very often, the dissonance between the objectives of entrepreneurial support programs for young people and the "hidden" objectives of the political players who have encouraged young people in their area to enroll in these programs is likely to encourage the failure of entrepreneurial support for young people with projects in agriculture. The psychological and above all cognitive profile of the young entrepreneurs who received support highlights the absence of morality, a lack of entrepreneurial culture, entrepreneurial spirit and perseverance, according to Raphaël Nkakleu et al. (2023), hence the interest of the present study, which justifies taking entrepreneurial culture into account for all downstream initiatives.

Through a learning process, entrepreneurial culture aims to improve the skills of entrepreneurs, especially young people and nowadays schoolchildren, and to give them easier access to the resources they need to carry out their projects when they come of age, precisely in the difficult context of underemployment in Burkina Faso. The entrepreneurial culture, particularly that taught to learners in rural areas, is becoming a serious alternative to be taken into account in the training and education of learners in general education establishments. The CEG de Bassemyam, which has implemented this initiative, is an illustration of this, even a pilot school.

*****

In short, the literature review enabled us to examine previous work in relation to the object of the research. The analysis carried out led to the evocation of three essential points: cooperation for development, promotion for agricultural entrepreneurship, agricultural entrepreneurship and rural development. These three points enabled us to discuss the writings and authors who have carried out research in the field of agricultural cooperation and the promotion of agricultural entrepreneurship. They have also highlighted the criticisms voiced by certain reluctant authors on the subject. Far from totally calling into question the idea of agricultural entrepreneurship, these are in fact safeguards which, if taken into account, would enable agricultural entrepreneurial initiatives to be led effectively towards conclusive or successful results.

PARTIAL CONCLUSION

The study of this first part, devoted to an apprehension of the keys and a review of the literature, enabled us to establish the theoretical framework by clarifying the scope of the essential notions of the subject and by mentioning the works on the research topic. In fact, the definitional approach we adopted removed any confusion as to our understanding of the concepts of cooperation, promotion, agricultural entrepreneurship and the rural commune. This made it possible to clarify our scientific orientation.

Furthermore, the literature review focused on writings related to development cooperation, the promotion of agricultural entrepreneurship, agricultural entrepreneurship and rural development, facilitated the orientation of our research and enabled us to situate our work in the continuity of our predecessors. In short, these two chapters have helped to orient our research in the direction of our quest by clarifying the essential points and orienting our research in relation to the writings of other authors.

PART TWO: RESEARCH METHODOLOGY AND ANALYSIS OF RESULTS

The second part of our research comprises two essential chapters. The first is devoted to research methodology. The second deals with the analysis of results. The research methodology will define the scope of the study, the survey population and sampling. We will also discuss the data collection method, the techniques and tools used to gather information from the target audience, and the data collection process.

The second chapter will analyze and interpret the results obtained from data collection with the target audience. In addition, we will test the hypotheses put forward at the start of our work. Finally, recommendations will be made to the Swiss Cooperation to enable it to improve its activities at the *Bassemyam* CEG.

CHAPTER III
RESEARCH METHODOLOGY

The research method is an indication of the approach we followed to carry out the field survey with the target audience. This study provides an insight into the Swiss Cooperation's contribution to promoting agricultural entrepreneurship at the CEG de *Bassemyam* in the rural commune of *Komsilga*. It covers the following points: scope of the study and survey population, sampling, data collection methods, data collection techniques and tools, and a description of the data collection process.

III.1. SCOPE OF THE STUDY AND DATA COLLECTION METHODS

This section presents the survey population, sampling and data collection techniques.

III.1.1. Scope of study

The delimitation of the field of study consists in segmenting the population in an optimal way in order to obtain a representation of the said population but in miniature. We adapted our survey to our study population, made up of resource persons connected with the project initiated by the Swiss Cooperation in the rural commune of *Komsilga*, and in particular at the CEG de *Bassemyam*. The field survey was carried out with this public (beneficiaries) in order to collect the information needed to carry out our reflection on the subject of promoting agricultural entrepreneurship in rural areas. The sampling is based on resource persons, people who can provide us with essential information on the subject. They are interviewed using an interview guide to obtain accurate and reliable information. Information that will facilitate the outcome of our analysis.

III.1.2. Data collection methods

The method identified for data collection is both qualitative and quantitative. The qualitative method meets the needs of our research on the contribution of Swiss cooperation to the promotion of agricultural entrepreneurship at the CEG de *Bassemyam* in the rural commune of *Komsilga*. The aim is to obtain quality information, not just quantity. Quality information allows us to obtain satisfactory and efficient results. The qualitative method involves questions that are more or less subjective, but which are formulated in an objective manner. It is more concerned with the quality of the information gathered than with the number of people who provide answers.

In other words, the qualitative method is one that enabled us, with the help of an interview guide, to gather the opinions of resource persons; in other words, to collect, apprehend and interpret information from reliable and qualified sources. Initially, we planned to interview three Swiss Cooperation resource persons directly involved in implementing the agricultural entrepreneurship project at the CEG de *Bassemyam*. In the end, two were able to respond favorably to our request for an interview, following the outline in the interview guide (see appendix for a copy). The expected answers were practically the same, which is why we no longer insisted on the third person, especially as another survey, this time quantitative, made up for this shortcoming as far as possible.

Awa Karambiri (2024, p.39) describes the quantitative approach as a research method using mathematical and statistical procedures to describe, explain and predict phenomena using historical data in measurable variable form. The aim is to deduce statistically measurable conclusions. Quantitative data collection is possible by means of a survey or questionnaire (see appendix for a sample). The questionnaire is used to collect quantifiable data. It takes the form of a series of questions asked in a precise order. It can also be used to gather a large number of testimonials or opinions. The main function of the questionnaire is to extend the survey and statistically verify the generalizability of the information and hypotheses previously established.

The questionnaire is therefore a series of questions methodically asked to define a case, a situation or a request from among a large number of possibilities, and to facilitate the gathering of testimonials. It's the ideal tool for gathering precise information from a large number of participants. It enables us to reach several people at once, and to collect quantifiable data that can be presented in table or graph form. Initially, we planned for a random probability sample of ten learners per class, of which there were four (6ème , 5ème , 4ème and 3ème), giving an initial total of forty learners to be surveyed, including twenty girls and twenty boys. In the end, thirty-two were able to take part in the survey, eighteen girls and fourteen boys. During the course of the survey, we noticed that the answers we received were practically all similar. The saturation point had thus been reached. We did not consider it necessary to continue the survey with the other eight learners on the initial sample list. We therefore chose a purposive sample

Notwithstanding this, a further twenty questionnaires were sent out to other members of the educational community (PTA, teaching and administrative staff, community leader). What techniques and tools were used in this study?

III.2. DATA COLLECTION TECHNIQUES AND TOOLS

The data collection techniques and tools we have used for our research are: documentary research, observation, interviews. These different techniques are essential to our research into the contribution of Swiss cooperation in promoting agricultural entrepreneurship at the CEG de *Bassemyam* in *Komsilga*.

III.2.1. Literature search

The documentary research led to the consultation of various documents. The documents consulted were mainly physical and digital works. Consulting them involved visiting libraries to find physical and digital works on the research question, as well as surfing the Internet to consult complementary documents: "The documents to be consulted are those related to the subject of study. These various readings helped to consolidate his field of knowledge around the subject of study" (D. Tibiri, 2023). The documents we consulted were books, articles, reports and dissertations relating to the contribution of cooperation and the promotion of agricultural entrepreneurship in rural areas. We also used the Swiss Cooperation information database for additional, precise and explicit information.

III.2.2. Participant observation

Participatory observation is a technique that enabled us to be in tune with reality on the ground. Immersing ourselves in the field of activities of the Swiss Cooperation through the CEG of *Bassemyam* was, for us, an experience of entrepreneurial, agricultural, professional and scientific awakening. The participant observation we carried out enabled us to analyze the attitudes, describe the actions and see the ways of proceeding of the trainers and instructors, as well as the pupils, teachers, parents and inhabitants of the rural commune of *Komsilga* who are involved in the agricultural project.

However, as Donald Long (2004, p. 2) points out, observation is more than a passive technique:

"It consists in systematically gathering qualitative and quantitative information. We observe individuals or groups. Before observing, the target behavior must be operationally defined. We can observe in two ways: by participating in the lives of the subjects we're observing, or by remaining withdrawn from their lives. We observe, so to speak, from inside or outside a group. Participatory observation is more demanding in many respects than disengaged observation. We try to build a more complex overall picture and a deeper understanding of the situation." .

In short, participant observation enabled us to immerse ourselves in the Swiss Cooperation's contribution to the promotion of agricultural entrepreneurship at the CEG de *Bassemyam* in the rural commune of *Komsilga*. It is also important to note that we are a teacher at the CEG de Bassemyam, and precisely the Director of the said establishment. This immediately implicates us in the project.

III.2.3. Interviews with resource persons

An interview is conducted using an interview guide and a questionnaire (see appendices XI and XIII). It (the interview) enables us to collect and analyze a number of elements, such as the attitude, opinion and feelings of the interviewee. It is a data-gathering technique which, unlike a questionnaire, establishes a special relationship between interviewer and interviewee. Interviews also serve to verify hypotheses and establish new ones. There are three types of interview: directive, semi-directive and non-directive or free. For this study, we opted for the free interview. With the help of an interview guide, we spoke with the resource people in charge of the project. The questionnaire was administered to learners and other members of the educational community. Using a mixed-method approach, the interview guide and questionnaire enabled us to gather the information we needed to understand our research work.

III.2.4. Data collection procedure

Data collection was carried out in two phases. The first phase consisted of interviews with resource persons. The second phase involved processing the data collected from the interviewees. The interview phase was conducted using an interview guide. This enabled us to talk to Swiss Cooperation staff, students involved in the project, parents and teachers involved in the project. In addition, we took a fortnight to complete this phase.

The various interview feedbacks enabled us to make the best possible use of the information gathered. For the data processing phase, we used the coding technique to process the data obtained. As a reminder, the interviews were semi-directive. We therefore noted the specific keywords in the responses. Finally, we entered the information collected into Google forms to capitalize on it and generate figures presenting quantitative elements for better analysis and explanation of the results.

*****

Finally, this third chapter enabled us to take a look at the institutional framework of our study through the presentation of the Swiss Cooperation and the CEG of *Bassemyam* in the rural Commune *of Komsilga*. In addition, we developed the methodological framework, defining the scope of the study and specifying the sampling. We also indicated the research method and data collection techniques and tools. Finally, we set out the process of data collection and processing.

CHAPTER IV
PRESENTATION OF RESULTS AND SUGGESTIONS

Analysis of the results is the first major point in this chapter. It enables us to examine and interpret the information gathered from the target audience. Next, we discuss the verification of hypotheses. This second point allows us to confirm or refute the hypotheses initially formulated. Finally, we conclude this chapter with recommendations for improving the agricultural entrepreneurship project at CEG *Bassemyam*, particularly in the school environment.

IV.1. INTERPRETATION AND DISCUSSION OF RESULTS

This main point will contribute to providing an analytical study of the information gathered from the resource persons in our field survey. To this end, we will proceed, on the one hand, to the interpretation of the results, and, on the other hand, to the discussion of the results obtained. This will enable us to further highlight our work.

IV.1.1. Results presentation

Interpreting the results enables us to comment on the data collected from the public. It involves highlighting and linking the information gathered through its interpretation in relation to the research topic.

Graph 1: Level of knowledge about agricultural entrepreneurship

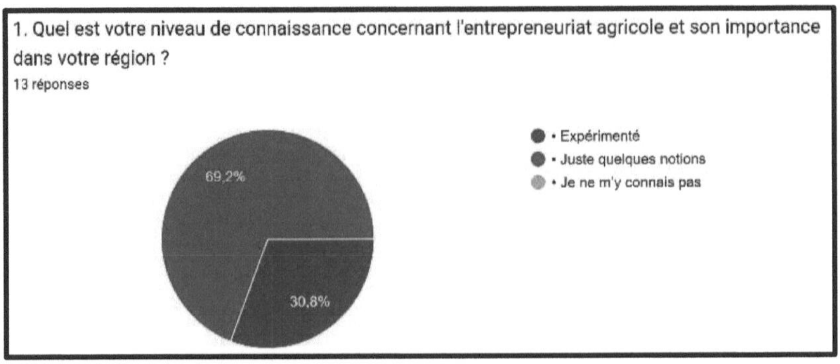

Source: Survey results from May 13 to June 05, 2024.

According to the data obtained in the first graph, 69.2% of our respondents claim to be moderately experienced in the field of agricultural entrepreneurship. In other words, these are people who, even if they can't claim to have mastered agricultural entrepreneurship to perfection, still have a few important notions to enable them to carry out certain actions. Furthermore, 30.8% claimed to have some expertise (experience) in the agricultural field. This helps to establish a certain complementarity between the various stakeholders. What is the level of apprehension about agricultural entrepreneurship among the respondents?

Chart 2: Level of participation in agricultural entrepreneurship programs

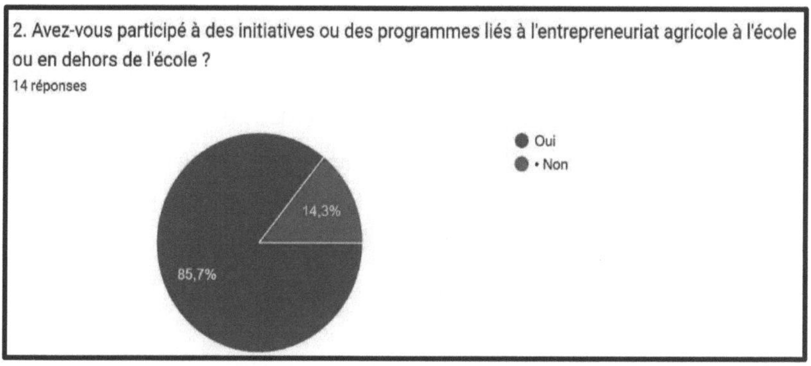

Source: Survey results from May 13 to June 05, 2024.

The second graph answers the question of whether our respondents had ever taken part in agricultural entrepreneurship initiatives or programs in or out of school. To this question, 85.7% answered "yes", while 14.3% answered "no". What then are the skills needed to succeed in this field?

Figure 3: Type of skills needed for agricultural entrepreneurship

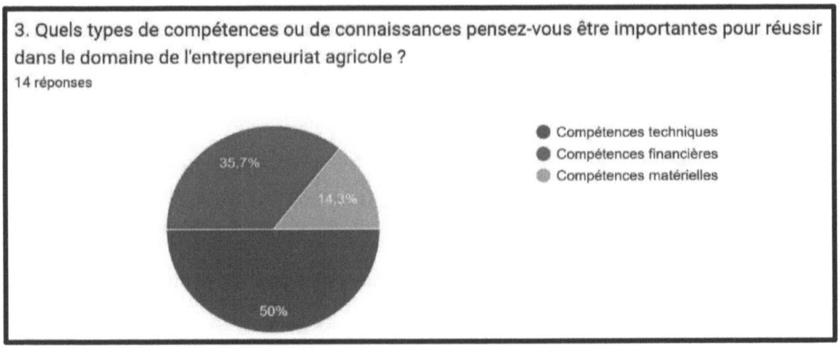

Source: Survey results from May 13 to June 05, 2024.

When asked what types of skills are needed to succeed in the field of agricultural entrepreneurship, 50% of our respondents raised the importance of technical skills; 33.5%, the vitality of financial skills; and 14.3% indicated that material skills are essential for success in agricultural entrepreneurship. But what might be the constraints to the success of this initiative? The following graph lends itself to this concern.

Figure 4: Challenges faced by *Bassemyam* farmers

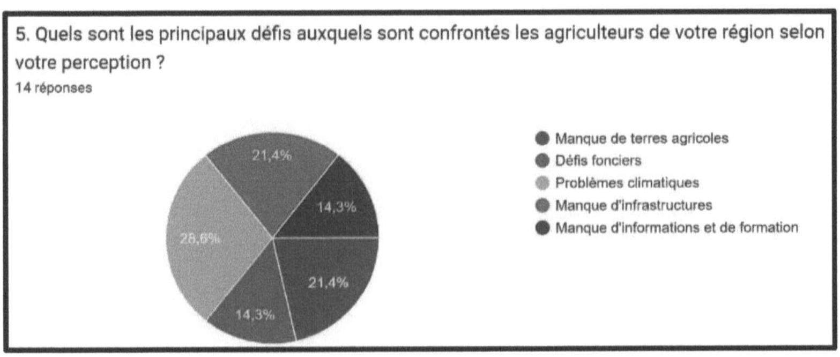

Source: Survey results from May 13 to June 05, 2024.

With regard to the challenges faced by farmers, 21.4% mentioned the lack of farmland and the lack of infrastructure respectively; 14.3% mentioned land tenure challenges and the same percentage mentioned the lack of information and training. In concrete terms, what was the nature of Swiss cooperation contributions?

Figure 5: Nature of cooperation contributions

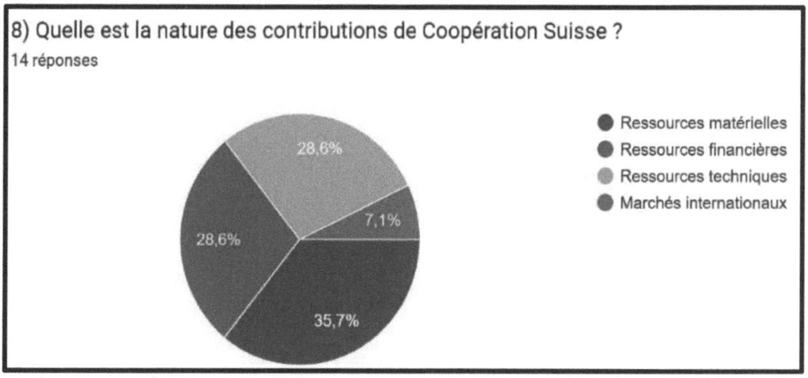

Source: Survey results from May 13 to June 05, 2024.

In addition, 28.6% mentioned climatic problems, notably the scarcity of rain and soil degradation. In addition, our surveys revealed that Swiss cooperation provides material resources (35.7%), financial resources (28.6%), technical resources (28.6%) and international markets (7.1%). What does the following graph say about the effectiveness of Swiss cooperation contributions?

Figure 6: Effectiveness of cooperation contributions

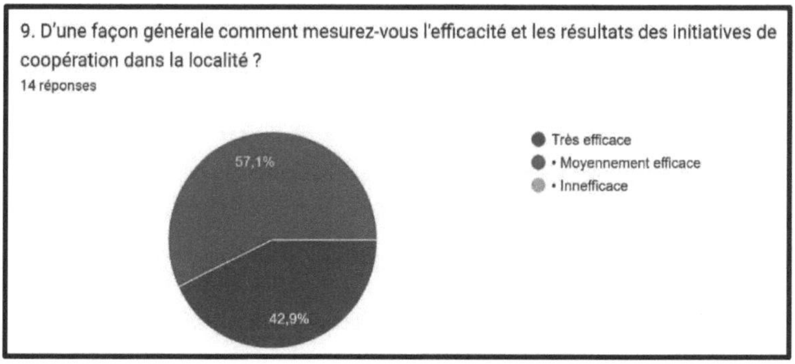

Source: Survey results from May 13 to June 05, 2024.

In graph 6, when asked how farmers rated the Swiss Cooperation's initiatives in the locality, 57.1% said they were moderately effective, while 42.9% said they were very effective. In this case, is it possible to evaluate the cooperation's intervention at the CEG in Bassemyam?

Figure 7: Evaluation of cooperation contributions

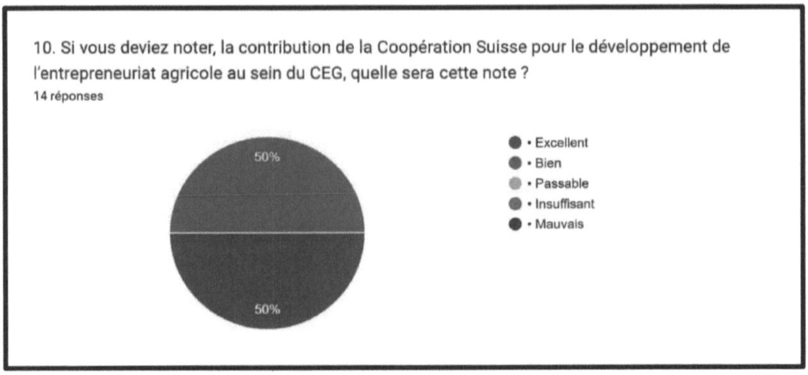

Source: Survey results from May 13 to June 05, 2024.

These results are almost identical to those shown in graph 7, where a balanced 50% of respondents said that the Swiss cooperation contribution deserved an Excellent rating, and the other said it deserved a Good rating. What assessments can be made of the above results?

Iv.1.2. Discussion of results

The discussion of results enables us to take an analytical look at the results obtained from the data collection with the resource persons. The discussion of results focuses, firstly, on an examination of the relationship between the respondents and agricultural entrepreneurship. Secondly, it examines the extent to which Swiss cooperation meets farmers' needs. Finally, we analyze the effectiveness of Swiss cooperation.

IV.1.2.1. Respondents' relationship with agricultural entrepreneurship

This first point in the discussion of results examines the level of knowledge and participation of respondents in programs related to agricultural entrepreneurship. Looking at the data obtained in the first graph, 69.2% of respondents consider themselves to be moderately experienced in the field of agricultural entrepreneurship. This means that these individuals, while not having a complete mastery of agricultural entrepreneurship, nevertheless possess significant knowledge enabling them to carry out certain actions in this field. At the same time, 30.8% of respondents claimed to have proven (experienced) expertise in the agricultural field. This distribution of experience levels suggests complementarity between the various stakeholders, where the less experienced can benefit from the expertise of the more experienced, thus fostering a knowledge-sharing and mentoring environment.

The second graph sheds light on respondents' participation in initiatives or programs related to agricultural entrepreneurship, whether in or out of school. It shows that 85.7% of respondents have already taken part in such initiatives, compared with 14.3% who said they had not. This high level of participation in educational and training programs in the field of agricultural entrepreneurship indicates a positive trend towards the acquisition of skills and the preparation of individuals for effective involvement in this sector.

The high proportion of participants in educational programs underlines the importance of these initiatives in developing the skills needed for agricultural entrepreneurship. It also reflects the interest and commitment of individuals to training and development in this field, despite varying levels of experience. The complementarity between the moderately experienced and the experts, coupled with a majority participation in training programs, creates fertile ground for the development of agricultural entrepreneurship.

Several conclusions can be drawn from these observations. Firstly, the diversity of experience levels among respondents highlights the need for training and mentoring strategies tailored to meet the specific needs of each group. Secondly, the high level of participation in training programs underlines the effectiveness of these initiatives and their crucial role in raising farmers' skills. Finally, this dynamic of complementarity and ongoing training can be seen as an essential lever for improving productivity and sustainability in the agricultural sector. By strengthening farmers' technical, financial and material skills through tailored programs, and by promoting the sharing of expertise, it is possible to create a more resilient and efficient agricultural ecosystem.

IV.1.2.2. Matching Swiss cooperation to farmers' needs

The second point in the discussion of the results relating to the study of Swiss Cooperation's contribution to farmers' needs enables us to give an opinion on the type of skills required in the field of study, as well as on the challenges faced by farmers and the nature of Swiss Cooperation's contributions. Indeed, when asked what skills are needed to succeed in the field of agricultural entrepreneurship, 50% of respondents stressed the importance of technical skills. A further 33.5% stressed the need for financial skills, while 14.3% said that material skills were essential for success in this field. This breakdown of responses highlights the diversity of skills perceived as essential for agricultural entrepreneurs, with a predominance of technical and financial aspects.

With regard to the challenges faced by farmers, the results show that 21.4% of respondents identified the lack of farmland and infrastructure as major obstacles. A further 14.3% mentioned land challenges, as well as lack of information and training. Climate issues were cited by 28.6% of respondents, who mentioned challenges such as rainfall scarcity and soil degradation. These data reveal a diverse range of difficulties faced by farmers, from material and land limitations to environmental and informational issues.

With regard to the contribution of Swiss Cooperation, the surveys revealed that 35.7% of respondents recognized the provision of material resources as a key support. A further 28.6% highlighted the provision of financial and technical resources. Finally, 7.1% mentioned access to international markets as a significant benefit provided by Swiss Cooperation. This distribution indicates a varied but substantial appreciation of the different types of support offered, with particular recognition for material and financial resources.

This information highlights not only the skills perceived as essential for agricultural entrepreneurship, but also the main challenges faced by farmers and the types of support provided by Swiss Cooperation. To maximize the effectiveness of its initiatives, it would be beneficial for Swiss Cooperation to continue diversifying its forms of support, taking into account the specific challenges and skills needs identified by farmers. This could include strengthening technical and financial training programs, improving infrastructure and promoting sustainable agricultural practices to meet climate challenges. By integrating these elements, Swiss Cooperation can not only consolidate its current impact, but also respond in a more targeted and effective way to the needs of the farming community.

IV.1.2.3. Effectiveness of Swiss Cooperation

The third point in the discussion of results concerns an analysis of the effectiveness of the cooperation. It allows us to see the successful and efficient aspects of the results, and to foresee possible perspectives. To this end, analysis of graph 6, on the evaluation of Swiss Cooperation initiatives by local farmers, shows that 57.1% of respondents considered these initiatives to be moderately effective. In contrast, 42.9% of farmers rated these initiatives as very effective. This distribution of opinions highlights an overall positive but nuanced perception of Swiss Cooperation's impact, with a relative majority seeming to recognize the effectiveness of the initiatives while pointing to room for improvement.

These results are largely comparable to those observed in Figure 7, which shows a balanced distribution of evaluations of Swiss Cooperation contributions. Indeed, 50% of farmers gave these contributions an "Excellent" rating, while the other half gave them a "Good" rating. This parity in responses indicates unanimous recognition of the value of Swiss Cooperation's contribution, while suggesting that certain aspects could be optimized to achieve the excellence perceived by all.

These data, taken from graphs 6 and 7, suggest a general recognition of Swiss Cooperation's efforts and their positive impact on the locality, although there are nuances in the degree of effectiveness and satisfaction. The ratings "moderately effective" and "good" represent a positive appreciation, while pointing to opportunities for continuous improvement. On the other hand, "very effective" and "excellent" ratings reflect a high degree of satisfaction and recognition of the successes already achieved.

Thus, for Swiss Cooperation, these results can serve as a basis for reinforcing the strengths identified by farmers, while addressing areas requiring adjustment. A more in-depth analysis of the reasons behind the moderately positive evaluations could provide valuable insights for increasing efficiency and overall satisfaction. In sum, these data reveal a broadly positive appreciation of the initiatives, accompanied by an invitation to pursue improvement efforts to achieve shared, universally recognized excellence.

IV.2. OPINION ON THE IMPACT OF THE PROJECT

This section highlights testimonials of the project's impact on the CEG and local communities in Bassemyam. To this end, we will relay testimonials from learners (the CEG de Bassemyam student delegate), the educational community (the staff delegate), the Parents' Association (the APE president), the village chief and community dean.

IV.2.1. For learners

The learners expressed themselves through the voice of the students' delegate: "I thank our teachers, the director, our parents and especially our Swiss friends because this initiative of the Swiss Cooperation has enabled us to :

- be assessed through a quarterly average entitled Production ;
- identify the consequences of climate change ;
- understand the need for tree planting (Photo plate no.1 in appendix XVII);
- learn about the various possibilities for reclaiming soil ;
- know the importance of the activities of our farming parents;
- to be able to help our parents in our spare time;
- to learn other trades such as poultry farming, gardening and commerce" (Photo plate n°1 in appendix XVII).

IV.2.2. Within the educational community

The staff representative said: "We are fully satisfied with the implementation of the Swiss Cooperation initiative, as it has enabled us to :

- revive children's interest in gardening and poultry farming;
- continue the collaboration between kindergarten, primary and post-primary students and the whole community around the agro-ecological poultry farm;
- strengthen collaboration between the colleges and other local establishments, with a view to sharing experience on agro-ecological poultry farms in the future;
- erase the image of a restrictive school with a more jovial, attractive and voluntary image thanks to gardening and animal husbandry activities.
- pay particular attention to learners in the 6ème and 5ème classes, in order to integrate environmental and entrepreneurial concerns into their mindsets at an early stage" (see

photo plate no. 2 in appendix XVIII).

IV.2.3. Parents' Association (APE)

The President of the Parents' Association (APE): "We'd like to thank the educational community, which initiated the project, and the Swiss Cooperation, which kindly co-financed it. We are very pleased that the activities have enabled us to :

- revive children's interest in farming and animal husbandry;
- help their parents with market gardening in the *Boulbi* lowlands;
- strengthen collaboration and solidarity at village level, with the involvement of parents, community leaders, local councillors and the Village Development Committee (CVD), to ensure the success of activities and the well-being of learners (see photo no. 6, page 48);
- have a positive impact on schools in the commune by implementing such initiatives for the real empowerment and independence of schools".

IV.2.4. At the level of the community elder representing the village chief

For the dean of the *Bassemyam* community: "The activities went well overall. We visited the Collège on several occasions to see how the work was progressing, and to advise the learners on their real role in the success of the activities (see photo plate n°2 in appendix XVIII).

In addition to raising awareness among parents in the village, we have not hesitated to recommend the project team to other resource people, so that they can benefit from technical and financial support in the face of the new challenges they face. That's why we don't hesitate to say:

Wend na yao yamb ti séké

La Kon Yam paanga ti ton song burkina karambissa faan!".

Translation: "God thanks you for thinking of us. May he give you the strength you need to help other students in Burkina Faso".

IV.3. VERIFICATION OF HYPOTHESES AND SUGGESTIONS

Verifying hypotheses and formulating suggestions will give us a better grasp of the object of our quest.

IV.3.1. Hypothesis testing

To carry out this verification process, we'll use a table to help explain our approach. (See page 47)

	ECONDARY IYPOTHESIS N°2	ECONDARY IYPOTHESIS N°1	IAIN HYPOTHESIS
IYPOTHESES	*Swiss cooperation has elped to diversify economic activities and improve living conditions for learners and local communities alike.*	*Swiss Cooperation has facilitated learners' access to ppropriate training, technical resources, market opportunities and revenues.*	*ctivities carried out by wiss Cooperation in the ural commune of omsilga have contributed) the promotion of gricultural ntrepreneurship at the 'EG de Bassemyam and to apacity-building for local uthorities*
ARGUMENTS	In graph 6, concerning the evaluation of Swiss Cooperation initiatives by local farmers, 57.1% of respondents rated these initiatives as moderately effective and 42.9% of armers rated them as very effective. This underlines an overall positive perception, with a relative majority seeming to ecognize the effectiveness of the initiatives. 50% of farmers gave these contributions an "Excellent" rating, while he other half gave them a 'Good" rating. This parity in responses indicates unanimous recognition of the value of Swiss Cooperation's contribution. **Our secondary hypothesis N°2 is thus confirmed.**	the surveys revealed that 35.7% of respondents recognized the provision of material resources as a key support. A further 28.6% highlighted the provision of financial and technical resources. Finally, 7.1% mentioned access to international markets as a ignificant benefit provided by Swiss Cooperation. This listribution indicates a varied ut substantial appreciation of he different types of support offered, with particular recognition for material and financial resources. **Secondary hypothesis 1 is therefore confirmed.**	n the sixth chart, 42.9% of farmers rated Swiss cooperation initiatives as very effective. In the seventh chart, 50% of farmers gave these contributions an "Excellent" rating, while he other half gave them a Good" rating. These data, aken from graphs 6 and 7, suggest a general recognition of Swiss Cooperation's efforts and heir positive impact on the locality. **The main hypothesis is confirmed.**

IV.3.2. Suggestions

The formulation of a recommendation is the contribution of ideas, suggestions with a view to bringing added value to an activity, to a thing. In this major point, we make recommendations to the Swiss Cooperation for the improvement of activities within the framework of the promotion of agricultural entrepreneurship at the CEG of *Bassemyam*. In this way, we are making our contribution to the optimization and effectiveness of its actions.

This is why we suggest that the State of Burkina Faso :

- Intensify technical and financial training **programs** for farmers to enhance their essential skills. Such training should be regular and accessible, covering aspects such as modern farming techniques and financial management. Mentoring and coaching sessions: Set up mentoring programs where more experienced farmers can guide novices, thus promoting the exchange of knowledge and the development of skills.

- **Promoting access to information and training:** Increase the availability and accessibility of information and training, by creating online platforms and distributing teaching materials tailored to farmers' needs.

- **Promoting sustainable agriculture:** Promoting sustainable agricultural practices to meet climate challenges, such as soil conservation, the use of resilient crops and efficient water management.

- **Support educational initiatives:** Continue to support and promote educational initiatives and training programs in agricultural entrepreneurship, both within and outside educational institutions. These programs are essential to effectively prepare farmers to meet the challenges of the sector.

To the Swiss Cooperation of :

- **Work on developing agricultural infrastructure:** Invest in building and improving agricultural infrastructure, such as irrigation systems, storage warehouses and rural roads, to improve farmers' working and production conditions.

- **Creating networks and communities:** Facilitate the creation of networks and communities of farmers to encourage the exchange of best practices, the sharing of experiences and mutual support;

- **Diversify sources of support:** Diversify types of support by introducing additional resources, such as consulting services, innovation subsidies and easier access to international markets;

- **Create climate resilience programs:** Develop climate resilience programs to help farmers adapt to changing conditions and mitigate the negative effects of climate change.

- **Evaluate and adjust resources:** Regularly evaluate the effectiveness of the material, financial and technical resources provided by Swiss Cooperation, and adapt these resources in line with farmers' feedback to better meet their needs.

- **Reduce the sabbatical period for grant renewal**: Continue to support banks with limited resources, such as schools.

- **Replicate the experience in other schools.**

*****

In short, this final chapter has enabled us to analyze and interpret the results obtained from processing the data collected from the target audience of the project to promote agricultural entrepreneurship in the rural commune of *Komsilga*, specifically at the CEG de *Bassemyam*. In addition, we were able to verify the hypotheses put forward at the outset of our report. Finally, recommendations were formulated to amplify the project and improve the contribution of Swiss Cooperation in *Bassemyam*, particularly at the Collège d'Enseignement Général.

PARTIAL CONCLUSION

The analysis of this second part focused on research methodology and results analysis. These two chapters enabled us to gain a better understanding of our research subject, to define our quest and to come up with efficient results. Indeed, the presentation of the Swiss Cooperation in Burkina Faso and its actions to promote agricultural entrepreneurship in the CEG of *Bassemyam* enabled us to better understand its involvement and impact in the rural commune of *Komsilga*.

In addition, the scientific nature of our research was indicated by the precision of the research methodology: scope of the study, sampling, method, data collection techniques and tools. In addition, analysis of the data collected from the target audience facilitated interpretation of the results. In this respect, the interpretation of the results led to the verification of the hypotheses put forward at the outset of our study. Finally, recommendations were formulated to help develop Swiss Cooperation's actions in the rural commune of *Komsilga*, particularly at the CEG de *Bassemyam*.

GENERAL CONCLUSION

The study conducted on the Swiss Cooperation's contribution to the promotion of agricultural entrepreneurship at the *Bassemyam* CEG in the rural commune of *Komsilga* enabled us to analyze the contribution of cooperation to the edification and empowerment of a rural population in the agricultural sector. Indeed, agricultural entrepreneurship seems to be an alternative for the economic and social development of rural populations. What's more, entrepreneurship in agriculture helps to combat youth unemployment. By training students from the rural commune of *Komsilga* through the CEG de *Bassemyam*, Swiss Cooperation is helping to train and build up young Burkinabè (A. Zongo, 2022).

Our objective in this study was to assess the effectiveness of Swiss Cooperation in promoting agricultural entrepreneurship in the rural commune of *Komsilga*, particularly at the CEG de *Bassemyam*. We also wanted to identify the specific measures taken by Swiss Cooperation to support agricultural entrepreneurship in the rural commune of *Komsilga*. In the same vein, we wanted to examine the impact of Swiss Cooperation results on the entrepreneurial skills acquired by local communities, as well as on their economic and social well-being.

The analysis carried out revealed that agricultural entrepreneurship is becoming increasingly important in the fight against poverty and youth unemployment. By introducing students to the subject, particularly in rural schools, they are better able to assimilate techniques, experiment with practice through the lessons they have learned, and exploit the skills they have acquired. In this way, young people and local communities can improve their skills, gain a better understanding of the rudiments of agricultural practice, modernize their techniques and increase their incomes.

We have also seen that collaboration with the Swiss Cooperation is part of a dynamic of support for local authorities. It contributes to their support within the framework of a national support process, by impacting the environment and living conditions of the population. Agricultural entrepreneurship at the CEG de *Bassemyam* sounds like the start of a beneficial practice from an economic, social and even technical point of view. An analysis of the results gathered from the target audience shows that agricultural entrepreneurship contributes to the development of the local population, boosting the agricultural sector and increasing production and incomes.

The hypotheses formulated were verified. This verification confirmed our position on the benefits of promoting agricultural entrepreneurship in rural areas, particularly in *Bassemyam*, at the Collège d'Enseignement Général. In addition, the verification of our hypotheses highlighted the contribution of Swiss Cooperation in supporting the Burkinabe population and, beyond, Burkina Faso as a whole. This study has edified us both in terms of the assimilation of agricultural techniques and practices, but also in the management of school prerogatives and in our role as Director of the CEG de *Bassemyam*.

In short, the analysis of Swiss Cooperation's contribution to promoting agricultural entrepreneurship at the Collège d'Enseignement Général de *Bassemyam* in the rural commune of *Komsilga* is a contribution to agricultural and financial education in schools. This project is a revolution in academic practice and a godsend in a context of security crisis. It provides and sustains hope, keeps the population busy and, in turn, highlights the socio-economic potential and agricultural assets of the rural commune of *Komsilga*.

BIBLIOGRAPHY

ARTICLES

- ALONSO J. A. and GLENNIE J. (2015). "What is development cooperation?" *Development Cooperation Forum Policy Notes - 2016*, ECOSOC/United Nations Economic and Social Council.
- BAKOUAN S. et al. (2020). "Entrepreneurship in Burkina Faso: a ray of hope for an increasingly ambitious youth" in *Développement économique et emploi en Afrique francophone, L'entrepreneuriat comme moyen de réalisation*, Observatoire de la Francophonie Economique (OFE), Presse de l'Université de Montréal.
- BOUTARY M. et al, (2014). "Accompagner l'entrepreneur dirigeant de PME: faut-il adapter les savoir sou les relations?" *Revue internationale PME*, 27 (3).
- BROWAEYS X. and CHATELAIN P. (2011). "Commune rurale ou commune urbaine: espace d'appartenance et catégorie de référence", in *Étudier une commune. Paysages, territoires, populations, sociétés*, Paris, Armand Colin, "Collection U".
- FITZ K. et al. (2017). "Entrepreneurship in the agricultural sector: A literature review and future research opportunities," in *Entrepreneurship Theory and Practices*, vol. 41, n° 1.
- JOUHARI N. (2017). "Entrepreneurship: what is it?" in *ICI Formation.*
- JUUKO B. (2022). "Agricultural entrepreneurship: some keys to developing the potential of Africa's rural areas" in *Entreprenante Afrique,* Thinking and acting for entrepreneurship in Africa.

REPORTS AND BRIEFS

- INSD (2019). "Fichier des localités du 5e RGPH".
- INSEE (2022). l'Institut National de la Statistique et des Etudes Economiques, "Commune rurale".
- KARAMBIRI A. (2024). *Prévention et gestion des conflits de travail à l'Institut d'Education et de Formation Professionnelle (INEFPRO) à Ouagadougou (Burkina Faso)*, Professional Master's thesis, Institut Africain des Industries Culturelles, 2024.
- OUATTARA D. (2020). *La coopération suisse au développement au Burkina Faso dans le domaine de la formation professionnelle de 1996-2016*, Master's thesis, Université Ouaga1, Pr Joseph Ki-Zerbo, Département Histoire et Archéologie.
- OUEDRAOGO M. et al (2009), *Les capacités des communes rurales au Burkina Faso*

: Naviguer entre l'apprentissage et le pré-requis, Groupe de Recherche Action sur le Foncier (GRAF), Réseau Décentralisation Burkina (RDB) and the Royal Tropical Institute (KIT), 2009.

- PPEA (2015), "Socio-economic transformation of rural areas in Benin", *Activity report to December 31, 2014.*
- ZONGO A. (2021), Rapport opérationnel et financier des activités du CEG de Bassemyam.

WORKS

- DUDDAL (2023), *Guide du formateur, Entrepreneuriat agricole*, 2023.
- LEGER J. et al (2016), *Open book Entrepreneuriat*, Paris, Dunod.
- NKAKLEU R. et al. (2023). *Understanding failure factors in agricultural entrepreneurial support. The case of young necessity entrepreneurs in Cameroon*, XXXIInd AIMS conference.
- TIBIRI D. (2023), *Techniques d'expressions écrites,* Editions DAMEL, 2023.

WEBOGRAPHY

- Local news (2022), *Promoting agricultural entrepreneurship and financial education in milieu schools: experience of learners at CEG de Bassemyam.* https://www.eda.admin.ch/countries/burkinafaso/en/home/news/news.html/content/countries/burkina-faso/en/meta/news/2022/6/Promotion-de-l-entreprenariat-agricole-et-l- education-financiere-en-milieu-scolaire accessed July 5, 2024.
- CNRTL (2012), Centre National des Ressources Textuelles et Lexicales "Definition of cooperation". https://www.cnrtl.fr/lexicographie/coopération accessed June 28, 2024.
- CNS-BF (2009), Conseil National de la Statistique du Burkina Faso, "Commune rurale". http://cns.bf/spip.php?article186 consulted on June 28, 2024.
- SDC (2021), *Programme de Coopération suisse au Burkina Faso 2021-2025,* Berne. https://www.eda.admin.ch/deza/fr/home/pays/burkina-faso.html and https://www.eda.admin.ch/countries/burkina-faso/fr/home/representations/bureau-cooperation.html?g=0587378cb8974dde7700e94ec3f90e86 accessed July 5, 2024.
- Fiches Pratiques Chef d'Entreprise (2016). "Comment promouvoir son entreprise". https://fiches-pratiques.chefdentreprise.com/Thematique/marketing-vente 1052/Guide/Comment-promouvoir-son-entreprise-305113.htm accessed June 28, 2024.
- HELVETAS Burkina Faso, (2024), "Promouvoir l'entrepreneuriat agricole". https://www.helvetas.org/fr/burkinafaso/ce-qu-on-fait/comment-on-travaille/nos-

projets/afrique/burkina-faso/BurkinaFaso_PAPEA accessed June 28, 2024.IFFAfrique (2024), "Entrepreneuriat agricole". http://iffafrique.org/entrepreneuriat- agricole/ accessed June 28, 2024.

- LONG Donald (2004), *Introduction à la recherche : les techniques de recherche,* Centre de Recherche et de Développement en éducation (CRDE), Université de Moncton, N.B., Canada E1A 3E9. Canada E1A 3E9, (506) 858-4886, 2004. https://fr.scribd.com/document/494851456/Module-1 consulted on June 30, 2024.
- ILO (2022). "ILO mobilizes to promote agricultural entrepreneurship", 2022. https://www.ilo.org/fr/resource/news/loit-se-mobilise-pour-promouvoir-lentrepreneuriat-agricultural accessed June 28, 2024.
- Sol-Vert (2022), Femme unies pour un envol vert, *Rapport Suivi et évaluation (environmental education).*
- SumUp "Entrepreneurship - What is entrepreneurship? entrepreneurship ? ", 2024. https://www.sumup.com/fr-fr/factures/termes-comptables/entrepreneuriat/ consulted on June 28, 2024.
- University of Peace (2018), "Seven steps to developing cooperation", 2018. https://www.universitedepaix.org/7-etapes-pour-developper-la-cooperation-dans-un- group accessed June 28, 2024.

Organization chart of Switzerland in Burkina Faso

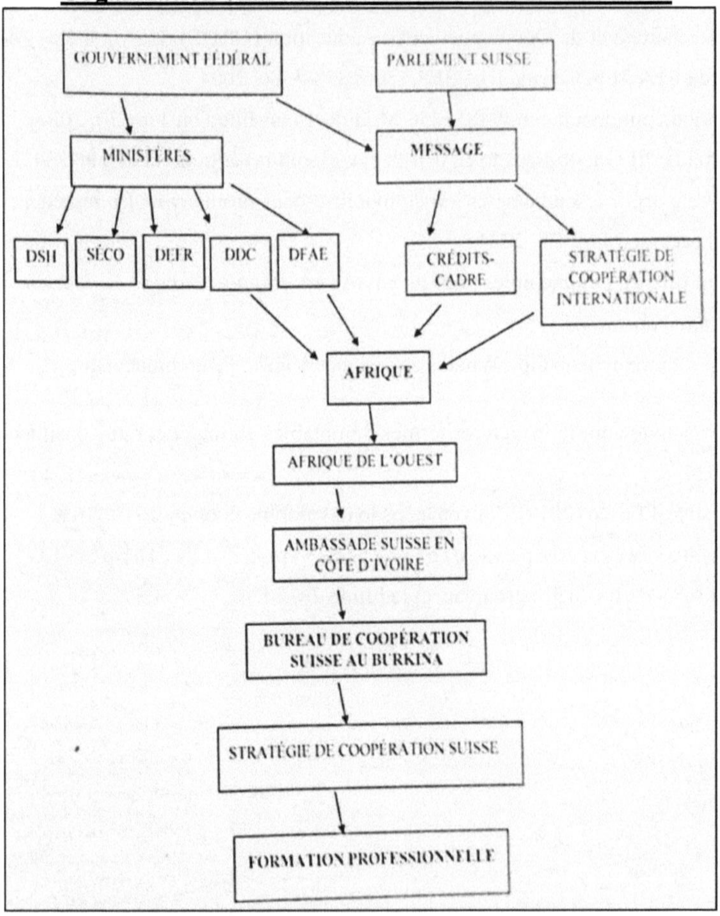

Source: Daouda Ouattara (2020)

Questionnaire

Hello everyone, I'm sending you this questionnaire as part of my dissertation for a professional Master's degree in Project Management. The subject of my dissertation ***is: "The contribution of Swiss cooperation to the promotion of agricultural entrepreneurship in the CEG of Bassemyam (Komsilga rural commune)".*** We would like to point out that this questionnaire is anonymous. We only wish to gather information on the contribution of Swiss cooperation in the promotion ofagricultural entrepreneurship. We would like to thank you for your understanding, your availability, the time you would take to fill in the questionnaire and, last but not least, for the sincerity of your answers.

Status indication: ....

---

1) How much do you know a b o u t agricultural entrepreneurship and its importance in your region?

- Experienced
- Just a few notions
- I don't know anything about it

2) Have you participated in any initiatives or programs related to agricultural entrepreneurship at school or outside school?

- Yes
- No

If so, which ones? ....

3) What types of skills or knowledge do you think are important for success in agricultural entrepreneurship?

- Technical skills
- Intellectual skills
- Physical skills

4) How do you think agricultural entrepreneurship could contribute to the development of your community?

....

6) What support or help do you think is needed to encourage more young people to get involved in agricultural entrepreneurship?

....

7) Does Swiss Cooperation provide you with this assistance?

- Yes
- No
- I don't know

8) What is the nature of Coopération Suisse's contributions?

- Material resources
- Financial resources
- Technical resources
- International markets

9) Generally speaking, how do you measure the effectiveness and results of local cooperation initiatives?

- Highly effective
- Moderately efficient
- Ineffective

Explain: ....

10) If you had to rate the Swiss Cooperation's contribution to the development of agricultural entrepreneurship within the CEG, how would you rate it?

- Excellent
- Fine
- Fair
- Insufficient
- Wrong

11) Do you have anything to add about Swiss Cooperation's contribution to the development of agricultural entrepreneurship at the CEG de Bassemyam?

....

Thank you for your cooperation

Interview guide

Hello Sir/Madam, this interview guide is part of the research I'm doing for my Master2 thesis.

The research topic in question is entitled: ***"The contribution of Swiss Cooperation in promoting agricultural entrepreneurship at the CEG de Bassemyam (Commune rurale de Komsilga)".*** The purpose of this interview guide is to enable us to obtain from you the information we need to carry out our research.

We would like to thank you for your understanding, your availability and the time you have given us for this interview.

Position held by respondent :
....

1) Type

- Woman
- Men

2) Your age range

- Between 18 and 30
- Between 31 and 40
- Over 40 years

3) Your length of service with Coopération Suisse

- Less than one (1) year
- Between one (1) and five (5) years
- Between six (6) and ten (10) years old
- More than ten (10) years

4) Can you describe your role within the company?

....

....

....

5) How long have you been working with *CEG Bassemyam*?

- Less than one (1) year
- Between one (1) and five (5) years
- Between six (6) and ten (10) years old
- More than ten (10) years

6) Could you describe in general the main obstacles facing farmers in Bassemyam?

- Barriers related to technical skills
- Barriers related to soil fertilization
- Obstacles related to adapted equipment
- Climate-related obstacles

Others:...

7) What specific programs or projects has Swiss cooperation implemented in this region to support agricultural entrepreneurship?

....

8) How do these programs contribute to COOPERATION SUISSE's economic and social development?

- By enabling them to
-
-
-

9) Have you provided appropriate technical training for learners at CEG *Bassemyam?* ?

- Yes
- No

10) If yes, how would you rate the effectiveness of this training?

- Highly effective
- Moderately efficient
- Unsuitable

Explain :

....

11) What kind of contributions do you offer farmers?

- Material resources
- Financial resources
- Technical resources
- International markets

Explain :

....

12) Do you think that your cooperation with the *Bassemyam* CEG is crucial to the success of their crops?

- Yes
- No
- I don't know

Explain ....

13) Generally speaking, how do you measure the effectiveness and results of your cooperation initiatives at the CEG de *Bassemyam*?

- Highly effective
- Moderately efficient
- Ineffective

Explain :

....

14) What are the Swiss Cooperation's future prospects for supporting agricultural entrepreneurship at the CEG de *Bassemyam*?

....

....

....

15) If you had to grade the Swiss Cooperation's cooperation with the beneficiaries of the CEG de Bassemyam in terms of implications for both sides, what would the grade be?

- Excellent
- Fine
- Fair

- Insufficient
- Wrong

16) What recommendations do you have for improving results between Swiss cooperation and the *Bassemyam* CEG?

....

....

....

17) What do you have to say about the issue of Swiss cooperation's contribution to promoting agricultural entrepreneurship at the CEG de *Bassemyam?*
?

....

....

....

Thank you for your cooperation

Photo 1: Gardening activities

Learners at CEG *Bassemyam* pay close attention to instructions

Student delegate and her classmates tending their watermelon and lettuce beds

Learners practicing planting plans

Source: OUIBGA Ibrahima, December 2023.

Photo 2: Members of the educational community and local players

The CEG staff and some local players from *Bassemyam*

Staff representative on the left, Chairman of the CVD and Chairman of the APE on the right

The village dean (bottom left), the local councillors (above), the VDC chairman (next to the chief), a pupil (next to the VDC chairman) and the CEG principal (bottom, far right of photo) plant the "community tree".

Source: OUIBGA Ibrahima, December 2023.

Map 2: Climatic conditions in *e Bassemyam* in the 1990s

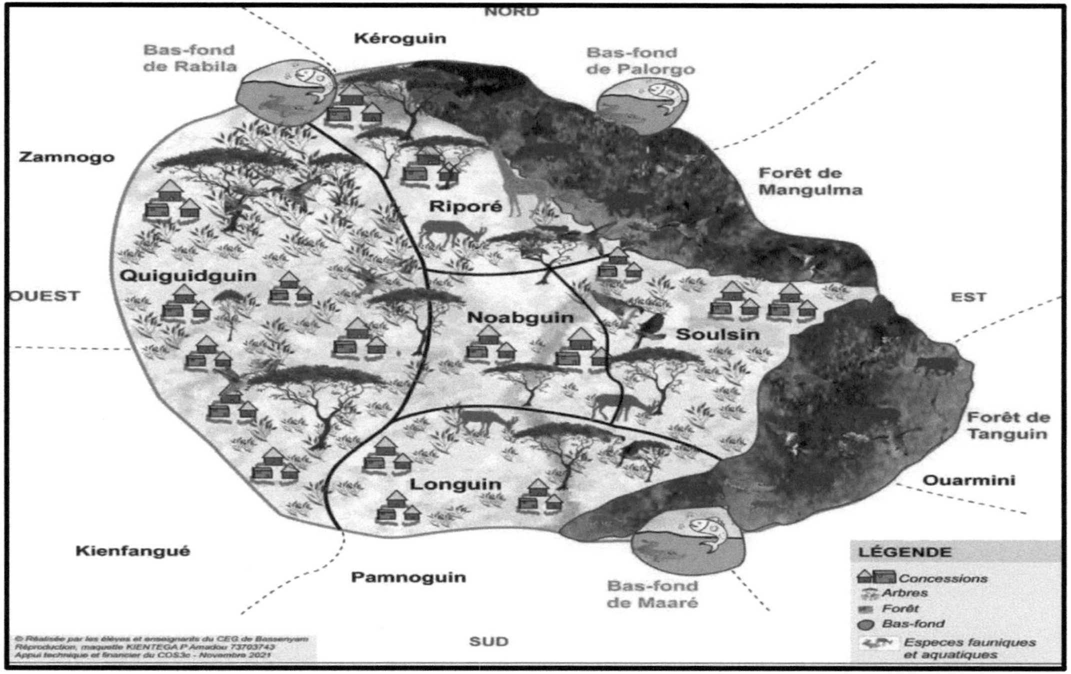

<u>Map 3</u>: Climate conditions in *Bassemyam* in 2021

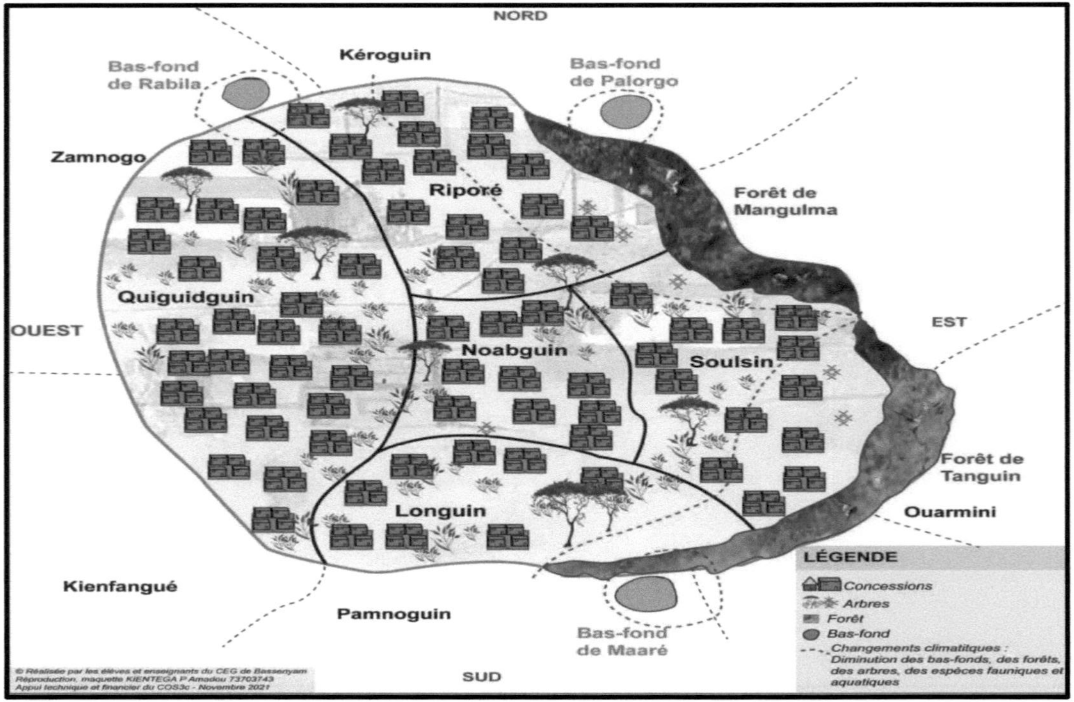

More Books!
yes

MIX
Papier aus verantwortungsvollen Quellen
Paper from responsible sources
FSC® C105338

Printed by Books on Demand GmbH, Norderstedt / Germany